We Are the Solution to Plastic Pollution
· A COLORING BOOK ·

copyright 2026 Imaginative Content, LLC
Santa Ynez, California
www.PlastiKong.com

Imaginative Content, LLC

*Plasti Kong trademark registration pending

C 2025 Gary Robinson
All Rights Reserved. Published 2026

ISBN 979-8-9939355-4-6

Imaginative Content, LLC
P.O. Box 1123
Santa Ynez, CA 93460
www.Imaginative-Content.com

> THIS PUBLICATION IS PART OF THE
>
> **PLASTIC POLLUTION AWARENESS PROJECT**
>
> PRESENTED BY IMAGINATIVE CONTENT

Check out www.PlastiKong.com for fun facts about plastic pollution
Get "The Rise of Plasti Kong" Game for iOS & Android

PLASTIC POLLUTION IS A REAL MONSTER

What is plastic pollution? It is the buildup of plastic trash like bags, bottles, and packaging in oceans and on land. Because it takes hundreds of years to break down, this waste hurts animals and sea creatures that eat or get tangled in it. Kids can help by reducing single-use plastic, recycling, and cleaning up litter.

Plastic Pollution is hiding everywhere! How many hidden PLASTIC POLLUTION monsters can you find on the pages of this coloring book?

Earth's Water Cycle
Clouds-Rain-Runoff-Oceans-Evaporation-Repeat

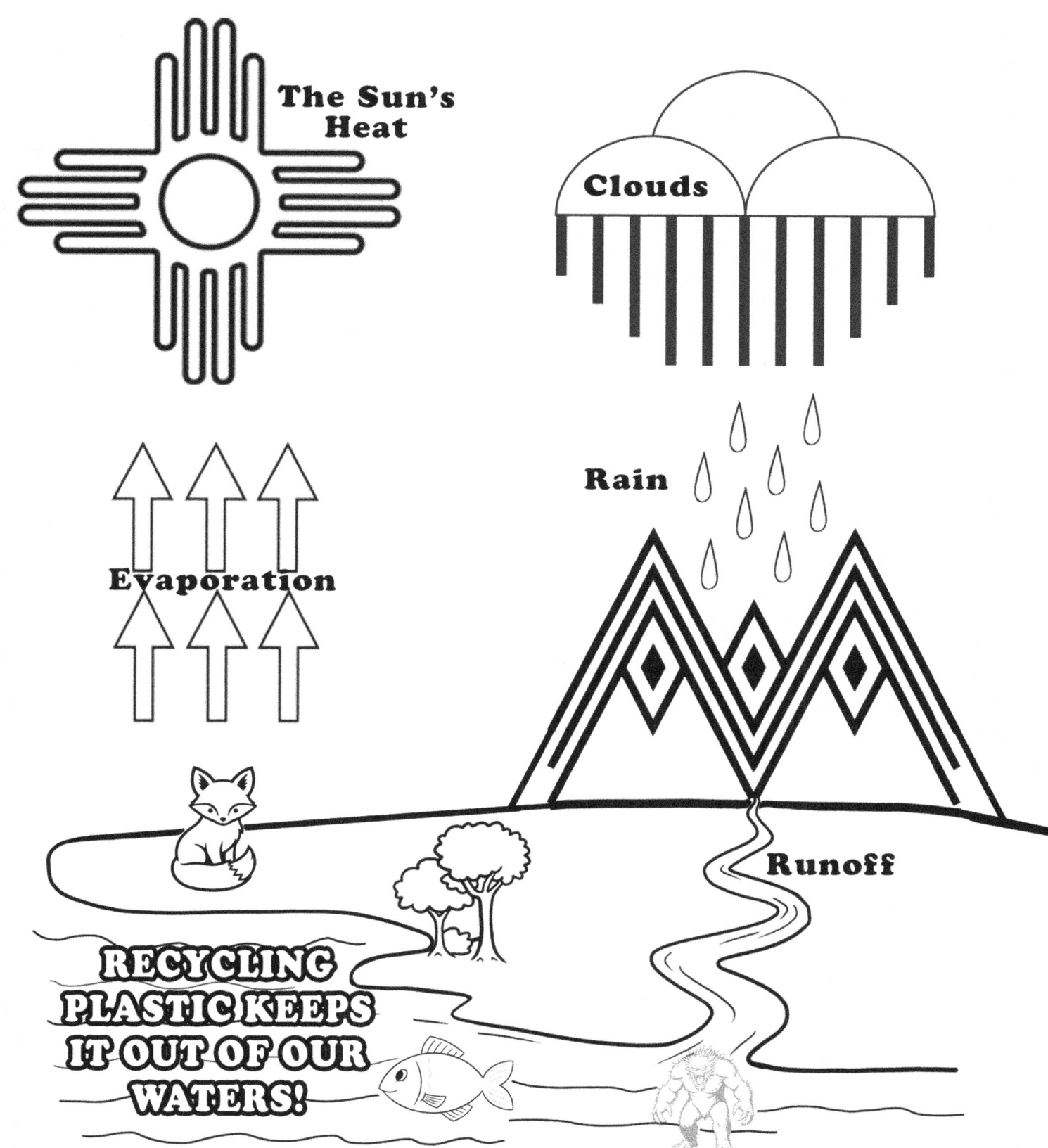

Help Build a World Without Harmful Plastic

Think of Ways You Can Help Reduce Plastic Pollution

JOIN THE CIRCLE OF EARTH PROTECTORS

Mother Earth's Children Thank You for Being an Earth Protector!

The Earth Protectors Badge of Honor

The Earth Protectors Pledge:
I promise to protect the earth from plastic pollution by:
1) Reducing the amount of plastic I use, 2) Recycling all plastic containers, and 3) Refusing single-use plastic containers whenever possible.

_____
your name here

DRAW YOUR OWN PICTURE of WHAT ENDING PLASTIC POLLUTION MIGHT LOOK LIKE

www.ingramcontent.com/pod-product-compliance
Lightning Source LLC
Chambersburg PA
CBHW080538030426
42337CB00023B/4788